Human Color Vision

Second Edition

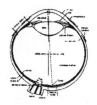

Peter K. Kaiser
and
Robert M. Boynton

OPTICAL SOCIETY OF AMERICA
WASHINGTON, DC

To Linda and Allie

Optical Society of America

ISBN 1-55752-461-0
Library of Congress Card Number 96-067204